NATIONAL
GEOGRAPHIC
KiDS

美国国家地理
双语阅读

Cheetahs

猎豹

懿海文化 编著

马鸣 译

第三级

外语教学与研究出版社
FOREIGN LANGUAGE TEACHING AND RESEARCH PRESS
北京 BEIJING

京权图字：01-2021-5130

图书在版编目 (CIP) 数据

猎豹：英文、汉文 / 懿海文化编著；马鸣译. —— 北京：外语教学与研究出版社，2021.11（2023.8 重印）
（美国国家地理双语阅读. 第三级）
书名原文：Cheetahs
ISBN 978-7-5213-3147-9

Ⅰ. ①猎… Ⅱ. ①懿… ②马… Ⅲ. ①豹－少儿读物－英、汉 Ⅳ. ①Q959.838-49

中国版本图书馆 CIP 数据核字 (2021) 第 228171 号

出 版 人　王　芳
策划编辑　许海峰　刘秀玲　姚　璐
责任编辑　姚　璐
责任校对　华　蕾
装帧设计　许　岚
出版发行　外语教学与研究出版社
社　　址　北京市西三环北路 19 号（100089）
网　　址　https://www.fltrp.com
印　　刷　天津海顺印业包装有限公司
开　　本　650×980　1/16
印　　张　37.5
版　　次　2022 年 3 月第 1 版 2023 年 8 月第 4 次印刷
书　　号　ISBN 978-7-5213-3147-9
定　　价　188.00 元（全 15 册）

如有图书采购需求，图书内容或印刷装订等问题，侵权、盗版书籍等线索，请拨打以下电话或关注官方服务号：
客服电话：400 898 7008
官方服务号：微信搜索并关注公众号"外研社官方服务号"
外研社购书网址：https://fltrp.tmall.com

物料号：331470001

记载人类文明
沟通世界文化
www.fltrp.com

Table of Contents

It's a Cheetah!

What runs so fast
it races by in a flash?

What looks
like it cries
but has no
tears in
its eyes?

What is covered in spots
and lives where it's hot?

4

It's a cheetah! (And we're not "lion.")

Cheetahs are large cats that look as cute and cuddly as a house cat. But you wouldn't want to snuggle up to a cheetah!

Cheetahs are powerful hunters with sharp claws and teeth.

Spotting Cheetahs

Cheetahs and leopards look alike because they both have spots. But they are different in many ways.

Cheetahs have "tear marks." These are black stripes that run from their eyes to their mouths. Leopards don't have stripes on their faces.

Cheetah   Leopard

 **T**
A
I
L

long and thin   short and thick

 **B**
O
D
Y

narrow   wide

 **H**
E
A
D

small   large

Safari Speedster

In a race
between a
lion, a
greyhound
dog, and a
cheetah,
which animal
would win?

The cheetah, hands down!

The cheetah is the fastest land animal on Earth. It can reach a running speed of 60 miles an hour in just three seconds. That's as fast as a sports car!

What makes a cheetah so fast?
Its body is built for speed.

A long tail balances the cheetah when it makes sudden, sharp turns.

A flexible spine helps it change direction quickly.

Long legs help it run fast.

A thin, lean body helps it move quickly.

Word Bite

PREY: An animal that is killed and eaten by another animal

Excellent eyesight makes spotting prey quick and easy.

Large nostrils let it breathe easily after running.

A small head makes the cheetah lighter.

Its claws don't completely pull back into its paws like other cats. The claws grip the ground when running, like cleats on a shoe.

Its deep chest makes breathing easier while running.

Great Hunters

camouflage
CAM-**oh-flaj**

Cheetahs are sneaky when they hunt! Their spotted coats act as camouflage in tall grass. They stalk their prey slowly and quietly.

When they get close, cheetahs chase their prey.

But cheetahs get tired quickly. Whew! They need to rest, too.

CAMOUFLAGE: An animal's natural color or shape that helps it hide from other animals

Word Bites

STALK: To move secretly toward something

How Cheetahs Live

Most cheetahs live in Africa. A very small number may still live in Iran, a country in Asia.

Like people, cheetahs can live in different habitats. Cheetahs live on the savanna and in areas with lots of plants. They also live on grasslands and in the mountains.

But cheetahs can't live near crowded buildings. They need open space.

Word Bites

HABITAT: The place where a plant or animal naturally lives

SAVANNA: A grassy plain with few trees in a hot, dry area

No matter where they live, male cheetahs stick together. Brothers live in a group called a coalition.

Female cheetahs live alone, except when caring for their cubs. Male and female cheetahs come together to have cubs. Then they live apart again.

coalition
koh-ah-LISH-un

17

Cubs

A mother cheetah has one to six cubs at one time. They are born blind and helpless.

But the cubs grow quickly! They can open their eyes and crawl in less than ten days.

The mother cheetah keeps the cubs safe in their den. If she needs to move, she carries them in her mouth.

Can you find the cheetah cubs?

The cubs' dark coats blend in with the shadows. The long, soft hair along their backs looks like the dry, dead grass.

The cubs are protected by camouflage. It's hard for predators to find them.

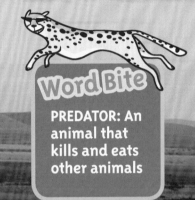

Word Bite

PREDATOR: An animal that kills and eats other animals

21

Playing Around

The cubs learn a lot from their brothers and sisters. They wrestle, stalk, and chase one another.

They practice skills they will need for hunting when they grow up.

When the cubs are older, the mother cheetah teaches them to hunt. She also shows the cubs which predators to avoid.

Royal Cats

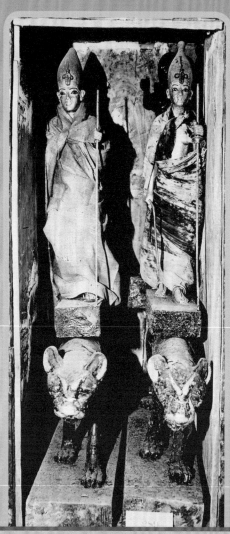

Statues of cheetahs from King Tut's tomb

Cheetahs have lived on Earth for a long, long time. As far back as ancient Egypt, pharaohs kept cheetahs as pets.

The famous pharoah King Tut was buried with many statues of cheetahs.

Some ancient Egyptians believed in a cat-goddess called Mafdet. They thought Mafdet could protect the pharaohs.

Art from ancient Egypt shows cheetahs on statues, furniture, and in paintings.

Mafdet
MAHF-**det**

Golden head of a cheetah found in King Tut's tomb

Cheetah Talk

Cheetahs make sounds that tell how they're feeling. Cheetahs can't roar like other big cats. But they can purr like a house cat.

Here's a dictionary for understanding cheetah talk:

Purring:
This is a low, motor-like sound, made when a cheetah is happy or content.

Bleating:
A cheetah bleats when it's upset. It sounds like a cat's meow.

Hissing:
When a cheetah feels angry or threatened, it may let out a sharp "h" sound.

Chirping:
Cheetahs chirp when they look for each other. The call sounds like a chirping bird.

Churring or stuttering:
During social meetings, cheetahs growl with a high pitch that stops and starts.

Growling:
A cheetah growls when it feels angry or threatened.

Saving Cheetahs

You need space to run, to jump, and to play—and so do cheetahs.

More people and more buildings push cheetahs onto smaller pieces of land. Cheetahs need lots of open space to live, to hunt, and to have babies.

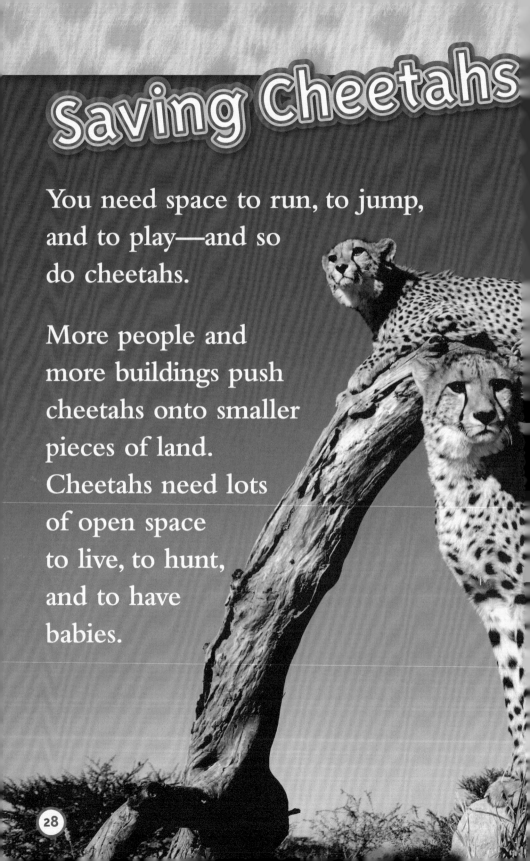

Less open space means cheetahs are disappearing. Today fewer than 7,100 cheetahs live in the wild.

But some people are working to save cheetahs. And we are learning more about these big cats.

The more we know, the better chance we have to keep cheetahs on Earth.

Glossary

CAMOUFLAGE: An animal's natural color or shape that helps it hide from other animals

HABITAT: The place where a plant or animal naturally lives

PREDATOR: An animal that kills and eats other animals

PREY: An animal that is killed and eaten by another animal

SAVANNA: A grassy plain with few trees in a hot, dry area

STALK: To move secretly toward something

▶ 第 4—5 页

是猎豹！

什么跑得特别快，一闪而过？
什么看起来在哭，但眼睛里却没有眼泪？
什么满身都是斑点，生活在炎热的地方？
是猎豹！（而且我们不是"狮子"。）
猎豹是大型猫科动物，看起来像家猫一样可爱，让人很想抱一抱。但是你应该不想依偎在猎豹的身旁！
猎豹是强大的猎手，有锋利的爪子和牙齿。

▶ 第 6—7 页

发现猎豹

猎豹和豹看起来非常像，因为它们都有斑点。但是它们有很多不同之处。
猎豹有"泪纹"。这些其实是黑色的条纹，从眼角一直延伸到嘴角。豹的脸上没有条纹。

第 8—9 页

极速狩猎者

狮子、灵猩和猎豹赛跑，哪种动物会赢？

猎豹，轻松获胜！

猎豹是地球上速度最快的陆生动物。它可以在仅仅 3 秒之内就将奔跑速度提升到每小时60 英里(约 96.56 千米)。这跟赛车一样快！

第 10—11 页

什么使猎豹跑得这么快？它的身体是为速度而生的。

大鼻孔让它在奔跑后更轻松地呼吸。

小小的脑袋让猎豹更轻盈。

厚厚的胸膛让它在奔跑时更轻松地呼吸。

极佳的视力让它又快又轻松地发现猎物。

它的爪子不像其他猫科动物那样完全缩到脚掌里。在奔跑时，它的爪子会像鞋子上的防滑钉一样抓住地面。

灵活的脊柱让它能快速改变方向。

瘦窄的身体让它能快速移动。

猎豹小词典

猎物：被另一只动物杀死并吃掉的动物

长长的尾巴让猎豹在急转弯时保持平衡。

长腿使它跑得飞快。

▶ 第 12—13 页

优秀的猎手

伪装

猎豹在捕猎时"鬼鬼祟祟的"！在高高的草丛中，它们那身长满斑点的毛成了伪装。它们慢慢地、悄悄地潜近猎物。

靠近之后，猎豹开始追逐它们的猎物。

但猎豹很快就会疲倦。呼！它们也需要休息。

猎豹小词典

伪装：动物天然的颜色或形状，帮助它不被别的动物发现

潜近：偷偷地接近某个东西

▶ 第 14—15 页

猎豹如何生活

大多数猎豹生活在非洲。很少一部分可能仍然生活在亚洲国家伊朗。

和人类一样，猎豹可以在不同的栖息地生活。猎豹生活在稀树草原和植被茂密的地区。它们也生活在草原和山区。

但是猎豹不能生活在建筑群附近。它们需要开阔的空间。

猎豹小词典

栖息地：植物或动物天然生长的地方

稀树草原：位于炎热干旱地区、树木稀少的草原

▶ 第 16—17 页

不管在哪里生活，雄性猎豹都会待在一起。兄弟们生活在被称作"豹群"的群体中。

除了照顾幼崽的时候，雌性猎豹都独自生活。雄性猎豹和雌性猎豹在一起孕育幼崽。然后它们就分开生活。

豹群

▶ 第18—19页

幼崽

猎豹妈妈一胎会生1到6只幼崽。它们在刚出生时看不到东西，非常无助。

但是幼崽长得非常快！不到10天，它们就能睁开眼睛，慢慢爬行。

猎豹妈妈把幼崽保护在兽穴里。如果要搬家，她就用嘴巴把它们叼走。

▶ 第20—21页

你能找到猎豹幼崽吗？

幼崽的深色皮毛和影子融为一体。它们背上的毛又长又软，看起来就像干枯的草。

幼崽通过伪装来保护自己。捕食者很难发现它们。

猎豹小词典

捕食者：杀死并吃掉其他动物的动物

▶ 第22—23页

四处玩耍

幼崽从它们的兄弟姐妹身上学到了很多。它们摔跤，追踪猎物，相互追逐。

它们练习长大后捕猎时所需要的技能。

当幼崽长大一些后，猎豹妈妈教它们捕猎。她也向幼崽展示要躲避哪些捕食者。

▶ 第 24—25 页

高贵的猫科动物

　　猎豹已经在地球上生活了很久很久。早在古埃及时期，法老就把猎豹当宠物养。

　　著名的法老图坦卡蒙与很多猎豹的雕塑埋葬在一起。

　　一些古埃及人信仰名叫"玛芙代特"的猫神。他们认为玛芙代特可以保护法老。

　　来自古埃及的雕塑、家具和绘画等艺术品上可以看到猎豹。

玛芙代特

图坦卡蒙墓中的猎豹雕塑

图坦卡蒙墓中的猎豹金首

▶ 第 26—27 页

猎豹的语言

　　猎豹用叫声表达自己的感受。猎豹不能像其他大型猫科动物那样吼叫。但它们可以像家猫那样发出呼噜声。

　　这里有一本词典，可以帮助理解猎豹的语言：

呼噜声:
这是一种低沉的、像马达一样的声音,猎豹在高兴或满足时会发出这种声音。

咩咩声:
猎豹在沮丧时会发出咩咩声。它听起来像猫的喵喵声。

咝咝声:
当猎豹感到愤怒或受到威胁时,它会发出尖锐的"h"的声音。

啁啾声:
猎豹在寻找同伴时会发出啁啾声。这种声音听起来像鸟的叫声。

颤鸣声或断断续续的咆哮声:
在社交会面时,猎豹会时断时续地、高音调地咆哮。

咆哮声:
当猎豹感到愤怒或受到威胁时,它会低声咆哮。

▶ 第 28—29 页

救救猎豹

你需要空间来奔跑、跳跃、玩耍——猎豹也需要。

越来越多的人和建筑物将猎豹挤到越来越小的土地上。猎豹需要大量的开阔空间来生活、捕猎、繁衍后代。

▶ 第 30—31 页

更少的开阔空间意味着猎豹在消失。如今,生活在野外的猎豹不足7,100 只。

但一些人在努力拯救猎豹。而且我们在了解更多关于这些大型猫科动物的知识。

我们了解的越多,我们就会有更好的机会保护地球上的猎豹。

词汇表

伪装：动物天然的颜色或形状，帮助它不被别的动物发现

栖息地：植物或动物天然生长的地方

捕食者：杀死并吃掉其他动物的动物

猎物：被另一只动物杀死并吃掉的动物

热带稀树草原：位于炎热干旱地区、树木稀少的草原

潜近：偷偷地接近某个东西